ANIMALS
Rebecca Woodbury, Ph.D., M.Ed.

Gravitas Publications Inc.

Animals

Illustrations: Janet Moneymaker

Copyright © 2024 by Rebecca Woodbury, Ph.D., M.Ed.

All rights reserved. No part of this publication may be reproduced, stored in a retrieval system, or transmitted, in any form or by any means, electronic, mechanical, photocopying, recording, or otherwise, without prior written permission from the publisher. No part of this book may be reproduced in any manner whatsoever without written permission.

Animals
ISBN 978-1-950415-56-4

Published by Gravitas Publications Inc.
Imprint: Real Science-4-Kids
www.gravitaspublications.com
www.realscience4kids.com

Photo credits: Cover & Title Page: By Eric Isselée, AdobeStock; Above, By kuritafsheen, AdobeStock; P.3. By kozorog, AdobeStock; P.5. By Leka, AdobeStock; P.6. By Happy monkey; P.7. By an, AdobeStock; P.9. By DBA, AdobeStock; P.10-11. By Eric Isselée, AdobeStock; P.12. By Leoniek, AdobeStock; P.13. By janstria, AdobeStock; P.14. By Silviu, AdobeStock; P.15. By vlad61_61, AdobeStock; P.16. By Vidu Gunaratna, AdobeStock; P.17. By kuritafsheen, AdobeStock; P.18. By sweetlaniko, AdobeStock; P.19. By dpep, AdobeStock; P.20. By New Africa, AdobeStock; P.21. By Alexey Kuznetsov, AdobeStock

There are many kinds of animals.

"Are cats, dogs, spiders, worms, and frogs all animals?"

"Yes. Even spiders and worms are animals."

Animals look different from each other.

Some animals have fur.

Some animals have scales.

Some animals have wings.

Animals have different

numbers of legs.

Animals live everywhere!

Some animals live in dirt.

Some animals live on top of lakes.

Some animals live in the ocean.

Some animals live in trees.

Some animals live in houses.

Sometimes I live in a house.

Other animals build their own houses.

How many different animals

can you see?

Many!

How to say science words

animal (AA-nuh-muhl)

fur (FUHR)

leg (LEG)

ocean (OH-shuhn)

scale (SKAYL)

science (SIY-ens)

wing (WING)

www.ingramcontent.com/pod-product-compliance
Lightning Source LLC
Chambersburg PA
CBHW041632040426
42446CB00022B/3487